Bradley's Manual on Growing and Curing Tobacco

by William L. Bradley

with an introduction by Roger Chambers

This work contains material that was originally published in 1864.

This publication was created and published for the public benefit, utilizing public funding and is within the Public Domain.

This edition is reprinted for educational purposes and in accordance with all applicable Federal Laws.

Introduction Copyright 2018 by Roger Chambers

IMPORTANT NOTE & DISCLAIMER

IMPORTANT NOTE :

As with all reprinted books of this age that are intended to perfectly reproduce the original edition, considerable pains and effort had to be undertaken to correct fading and sometimes outright damage to existing proofs of this title.

At times, this task can be quite monumental, requiring an almost total rebuilding of some pages from digital proofs of multiple copies. Despite this, imperfections still sometimes exist in the final proof and may detract slightly from the visual appearance of the text.

Some images may suffer from reduced quality due to anomalies in the original scan.

DISCLAIMER :

Due to the age of this book, some methods or practices may have been deemed unsafe or unacceptable in the interim years. In utilizing the information herein, you do so at your own risk.

We republish antiquarian books with no judgment or revisionism, solely for their historical and cultural importance, and for educational purposes.

Self Reliance Books

Get more historic titles on animal and stock breeding, gardening and old fashioned skills by visiting us at:

http://selfreliancebooks.blogspot.com/

Disclaimer

This book was written in an age when little was known about the ill effects of tobacco.

The material presented herein is intended to be strictly for educational purposes with the purpose of enlightening readers about the historical uses of tobacco. Publication of the material is neither an endorsement, nor a criticism of its contents. This book is presented as part of large series of educational material on the history and cultivation of tobacco.

As the reader, please consider it your duty to consult with a medical doctor before utilizing tobacco. It is also the reader's duty to become familiar with local, state, provincial and federal laws relating to the growing of tobacco.

As the author, publisher and retailer cannot control how the reader utilizes the historical information presented in the pages herein, they hereby disclaim any liability to any party for any loss, damage, disruption, death or other liability that may be incurred by the reader's misuse of this material.

introduction

Here at **Self-Reliance Books** we are dedicated to bringing you the best in *dusty-old-book-knowledge* to help you in your quest for self-sufficiency and independence.

We're so pleased to bring you this old title on growing tobacco crops. We republish there mainly for their historical value and for educational purposes, and it should be that some of the information is best looked at in the historical aspect, due to the obsolescence of some practices or methods.

This special edition of ***Bradley's Manual on Growing and Curing Tobacco*** was written by William L. Bradley, and first published in 1864, making it over 150 years old.

This super-short, fast read features sections on *Tobacco-Producing Countries and Total Production, Varieties Used in Manufacturing, Described, The Production of Tobacco in the States and Territories, The Time for Transplanting, Tobacco-Barns or Drying-Houses*, and more.

Another great old book and a must-read for all those interested in the historical aspect of the Tobacco Industry in the United States.

~ *Roger Chambers*
State of Jefferson, March 2018

CONTENTS.

	PAGE.
Note to the Reader,	3
Tobacco,	5
Tobacco-producing Countries, and total production,	6
Varieties used in Manufacturing, described,	6
The Production of Tobacco in the States and Territories,	7
Exportation and Importation,	8
Soil for Growing Tobacco, and its Preparation,	8
The Plant-bed,	8
The Time for Transplanting,	9
Clean Culture Essential,	10
Worming,	11
Topping,	12
Suckering,	12
Harvesting, mode of in Cuba, and in the Connecticut Valley,	13
Stripping and Piling,	16
Recapitulation,	17
Expense and Profit of Tobacco-Growing,	18
Tobacco-barns or Drying-houses,	20
Bill of Timber and Lumber for a Tobacco-barn,	21
A Model Tobacco-barn,	23
Consumption and Exportation of Tobacco,	24
Analyses of Tobacco Soils and Tobacco,	25
Table of Analyses,	28
Testimonials,	29 to 36
Directions for Using Bradley's Patent Tobacco Fertilizer,	36
Dr. Jackson's Letter,	3d page of cover.

NOTE. Any person engaged in, or about to engage in growing tobacco, desirous of obtaining a copy of BRADLEY'S TOBACCO GROWERS' MANUAL, can do so by sending his address to Wm. L. Bradley, 24 Broad St., Boston.

NOTE TO THE READER.

HAVING devoted much time during the last two years to making myself acquainted with the most approved modes of cultivating and curing tobacco, I became convinced of the necessity and importance of manufacturing a special fertilizer for the tobacco crop. Accordingly, I introduced last season (1863) a fertilizer known as Bradley's Patent Tobacco Fertilizer, which I was very confident would meet the wants of tobacco growers. The trial has fully met my expectations, as may be seen by reading the "Testimonials" at the end of this Manual, and therefore I purpose to continue to manufacture the article, being confident as I now am, that it will become a standard fertilizer for tobacco, not only as an aid to barn-yard and hog manure in starting the plants vigorously and promoting early maturity, with a largely increased crop, but that it may be successfully used where barn-yard manure cannot be had, as seen by the "Testimonials." The quality of this fertilizer will be fully equal to what it was last year. The price will vary according to the cost of the materials of which it is made.

I give a few of the statements of the Tobacco growers of the Connecticut Valley who used Bradley's Patent Tobacco Fertilizer the past season, to which the reader's attention is specially invited.

I also invite attention to that part of this Manual that treats of the most approved modes of growing and curing tobacco in New England, where its culture has been introduced, as drawn from the very best sources of information, to wit, experimental, observational, and conversational, the growers of tobacco having been consulted at their homes where they were visited. The contents, with the numerous cuts of this Manual, will serve greatly to aid beginners in entering upon the cultivation of tobacco, — a branch of husbandry exciting much interest at the present time in consequence of the extremely high prices which tobacco now commands.

I would also further invite attention to the chemical analyses of tobacco soils and plants from Maryland and Massachusetts, lately made by Charles T. Jackson, M. D., of Boston, at the request of the Commissioner of Patents.

WM. L. BRADLEY,

MANUFACTURER OF
- BRADLEY'S PATENT TOBACCO FERTILIZER,
- BRADLEY'S X L SUPERPHOSPHATE OF LIME,
- COE'S SUPERPHOSPHATE OF LIME.

24 BROAD STREET, BOSTON.

TOBACCO.

Tobacco, like maize, is indigenous to tropical America. Columbus was invited to smoke it by an aboriginal chief of Cuba, in 1492. Since then it has become a luxury of the rich, and a solace of the poor throughout the world. It was found in Virginia, toward the close of the 16th century, and soon after was introduced into England. Its cultivation by the English was begun in the "Old Dominion" in 1611, about a quarter of a century after its first discovery in that colony. The Indians had grown it and smoked it from time immemorial. Its cultivation rapidly increased. King James I. issued proclamations to restrain its use; but they were powerless, for the demand increased, and was greater than for any other product of the province. In 1617, it was sold from 37½ to 75 cents a pound. In 1621, every person was required to grow 1000 plants of 8 leaves, weighing 100 lbs. In 1622, there was grown in the Colony of Virginia, 60,000 lbs. Its culture spread into Maryland, the Carolinas, Georgia, and Louisiana. Its introduction into foreign countries encountered violent opposition. Its importation into Turkey was forbidden, and those guilty of smoking the weed were condemned to death. The Grand Duke of Moscow prohibited its introduction under the pain of the "knout," for the first offence, and of death for the second: in other parts of Russia, smoking was denounced, and smokers condemned to have their noses cut off. Pope Urban VIII.

anathematized all who smoked in churches. Yet, notwithstanding all this opposition, the weed was imported and smoked and chewed and snuffed by emperors and kings, Popes and cardinals, rich and poor, Christian and antichristian, as the history of the extensive exportations from the tobacco-growing States clearly shows. It is well known, that many of the sovereigns of the Old World derive large revenues from the tobacco-trade with the New. It will be noted that the opposition to growing tobacco and smoking it is no modern discovery; also, that this opposition is as powerless now as it was in olden times.

Next to the United States the principal tobacco-producing countries are some of the West India Islands, the States of Central and South America, Cuba, Hayti, Brazil, Peru, etc.; in the East Indies, Manila, Java, China, etc., Asia Minor, Egypt, Turkey, Greece, Hungary, the southern part of Russia, Holland, Belgium, the States of Germany, many of the departments of France, Algeria, Corsica, and Upper Savoy, are all noted for the culture of tobacco. The total production of the world is estimated as follows: Asia, 399,900,000 pounds; Europe, 281,844,500; America, 248,280,500; Africa, 24,300,000; Australia, 714,000, making in all 995,039,000 pounds.

The French manufacturers describe the varieties used in manufacturing as follows: the Virginian, strong, very aromatic, and highly valued for snuff; the Kentucky, strong, large leafed, very choice; the Maryland, light, odoriferous, large-leafed, and good for pipes; the Havana, unequalled for cigars; other tobaccos from the West Indies, Central and South America, are used for cheap cigars.

The production of tobacco, in the several States and territories for the years named, is given as follows in the United States census returns: —

STATES AND TERRITORIES.	1840. lbs.	1850. lbs.	1860. lbs.
Maine,	30		
New Hampshire,	115	50	21,000
Vermont,	585		12,000
Massachusetts,	64,955	138,246	3,233,000
Rhode Island,	317		1,000
Connecticut,	471,657	1,267,624	6,000,000
New York,	744	83,189	5,764,000
New Jersey,	1,922	310	149,000
Pennsylvania,	825,018	912,651	3,182,000
Delaware,	272		10,000
Maryland,	24,816,012	21,407,497	38,411,000
District of Columbia,	55,559	7,800	
Virginia,	75,347,106	56,803,227	123,968,000
North Carolina,	16,772,359	11,984,786	32,853,000
South Carolina,	51,519	74,285	114,000
Georgia,	162,894	423,924	919,000
Florida,	75,274	998,614	758,000
Alabama,	273,302	164,990	221,000
Mississippi,	83,471	49,960	128,000
Louisiana,	119,824	26,878	41,000
Texas,		66,897	98,000
Arkansas,	148,439	218,936	1,000,000
Tennessee,	29,550,432	20,148,932	38,931,000
Kentucky,	53,436,909	55,501,196	108,102,000
Ohio,	5,942,275	10,454,449	25,529,000
Michigan,	1,602	1,245	121,000
Indiana,	1,820,306	1,044,620	4,658,000
Illinois,	564,326	841,394	7,014,000
Wisconsin,	115	1,268	87,000
Minnesota,			38,000
Iowa,	8,076	6,041	313,000
Missouri,	9,067,913	17,113,784	26,435,000
Kansas,			17,000
California,		1,000	3,000
Oregon,		325	
Utah,		8,467	
New Mexico,		70	
Total,	219,163,319	199,752,655	428,121,000

The annual exports of the United States were as follows for the years named below: —

YEARS.	BALES.	CASES.	HHDS.	VALUE.
1855,	12,913	13,366	150,213	$14,712,468
1856,	17,772	9,384	116,962	12,221,843
1857,	14,432	5,631	156,848	29,662,772
1858,	12,640	4,841	127,670	17,009,767
1859,	19,651	7,188	198,840	21,074,038
1860,	17,817	15,035	167,274	15,906,547
Total,	95,225	55,445	917,807	$110,587,435

The importations of tobacco are chiefly from Cuba, and four-fifths of these are of cigars, valued at about $4,000,000 annually.

SOIL AND ITS PREPARATION.

In the cultivation of tobacco, the first thing to be considered is the soil and its preparation. A rich, sandy loam, capable of retaining moisture, is deemed best by the growers in Cuba. A soil somewhat similar is found best in this northern latitude. Hence the success of the growers of the weed in the valley of the Connecticut. Good cultivation of such a soil, with plenty of barn-yard manure, thoroughly incorporated with the soil, with Bradley's Tobacco Fertilizer, will make a tilth or plant-bed, that with care and culture will insure a good crop.

THE PLANT-BED.

Select a good spot; one newly cleared is preferable, where it can be had, because free from weeds and grass. In selecting a spot, a warm exposure should be chosen, not wet, but moist; work in a liberal amount of manure and wood ashes, to the depth of six inches, — hog-manure is preferred by some, — then rake in Bradley's Tobacco Fertilizer, and when all are well mixed and comminuted, or pulverized, it is ready for the seed. For every square rod to be sown, mix an even table-spoonful of seed with plaster, and sow it as even as possible; after sowing, roll or tread the bed; beds three feet in width are convenient for weeding and watering, if need be. The time for sowing will vary from the first to the middle of April, along the Connecticut valley, The weeding of the plant-bed must be timely done, while the weeds are very small, so as not to injure the plants in pulling the weeds.

As it is important to have the Connecticut seed-leaf variety, it may be obtained of Messrs. McELWAIN & BROTHER, of Springfield, STOCKWELL & SPAULDING, of Northampton, and GEO. A. ARMS & Co., of Greenfield.

THE TIME FOR TRANSPLANTING.

Be not in too great a hurry to set the plants: from the 10th or 15th to the 25th of June is deemed early enough to transplant. Paoli Lathrop, of South Hadley, would prefer the latter date for field-culture. He adds, "the plant requires strong, warm land, such as would produce forty bushels of corn per acre, with at least ten cords of good manure, 200 lbs. of plaster [500 lbs. of Bradley's Tobacco Fertilizer] per acre." Mr. Lathrop would mark the rows three and a half feet apart one way, [they should run north and south, so as to give the full effect of the sun], and set the plants two feet and four inches apart in the rows. Many tobacco growers are in too great a hurry to set their plants, and sometimes use them before they are sufficiently grown. They should remain in the bed till well-rooted, and have good sized leaves. While in the bed they are secure from cut-worms, the pests of young plants. Mr. Lathrop remarks, "If the plants are large, the leaves four or five inches in length, the worm may eat some nights, and not injure the plants unless the centre-bud is removed. When that is gone, re-set." If the weather be dry when the plants are set, watering may be deemed necessary. It is recommended to take the plants from the bed by the use of some pointed implement. Mr. Beardslee, of Connecticut, recommends the covering of the plants with fresh-mown grass when they are first set, if the weather is sunny and warm. In a week or so, it may be removed. Mr. Crafts of Whately, recommends hay run through the cutting machine. It would be well in transplanting to use Bradley's Tobacco Fertilizer in the place where the plants are set, so as to give them a quick and vigorous start, as this is very essential to one who would raise a good, well-matured crop. So long as it will do to re-set, the grower should daily go through his field and fill the places made vacant by worms. Below may be seen a plant as it should be set, and as it is when set.

CLEAN CULTURE ESSENTIAL.

From the time the plants are set, the weeds should be eradicated, and the worms destroyed, such as cut the stems, and gnaw the leaves when the plants are first set out. They may be found just below the surface, near the injured plants.

Clean culture must be strictly adhered to by all who would succeed in growing a remunerating crop. A successful grower in the valley of the Connecticut, recommends that, in cultivating the growing crop of tobacco, the weeds should be cut up without stirring the soil much below the surface, in order to avoid injuring the rootlets that feed the plants from near the earth's surface. A light hoe, such as can be made of a saw-plate, and ground to an edge, and kept sharp, is a fit implement to destroy weeds among the growing plants of tobacco. A week after the plants are set, the cultivator may be run between the rows. After the plant attains much size, such a hoe as described, and the fingers, are the most useful implements for weeding, worming, and topping the plants. "*Clean Culture*," is the motto of every successful tobacco grower,— especially is it so on all lands that have been manured from the barn-yard, and cropped for many years.

WORMING.

The plants must be carefully watched from the time of setting to the time of harvesting, to protect them from the ravages of the cut-worm and the tobacco-worm, the latter being the larva of the *Sphinx Carolina*, the moth being of a gray color. The worm is large when grown, of green color, with a caudal horn-like appendage, and is a very disgusting looking creature. It requires the greatest vigilance to preserve the plants from being injured by tobacco-worms. Morning and evening, should the plants be examined, and both the worms and the eggs of the moth should be destroyed by the thumb and fingers. Be vigilant in worming, or else many of the best leaves of the plants will be ruined for wrappers. This may be done by faithful boys, girls, and women.

TOPPING.

A Connecticut grower says, "Top when the majority of the plants are ready to blossom, leaving ten or twelve leaves below." Another says, "Let as many of the plants blossom as possible without forming seed. Then break off to a good leaf. There is more danger of topping too high than too low." A Cuban grower says, "When the plants have from twelve to fourteen good leaves, and are about knee-high, begin to top by nipping off the bud with the thumb and finger, taking care not to injure the leaves near the place of topping; for in a good season the top leaves will grow nearly as large and ripen as soon as the lower ones." The cut shows a plant ready to top. The general average in Cuba is from twelve to fourteen leaves to a plant, and the writer adds that when "from sixteen to eighteen can be obtained," it is desirable to do so. Those who engage in topping should immediately, when they pause in the work, wash their hands, as the acrid juice of the plant will produce soreness of thumb and fingers. The cut below shows a plant nearly mature.

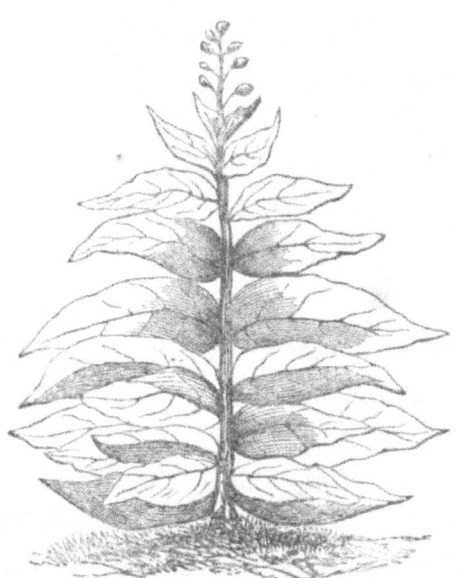

SUCKERING.

Suckers, which may start from every leaf, should be removed from the plant as soon as they appear after topping. Like worming, it must be promptly and faithfully done, and may be done in connection with worming, provided both can be equally well done at the same visitation. It would be quite impossible for all employees to do both equally well at the same time. Let the grower dictate as to this matter according to his experience. So important was suckering formerly regarded in Virginia, and so surely is it that suckers injure the quality of the tobacco, that at one time penal laws were enacted to prevent negligence in this respect. Below is a cut of a plant that needs suckering.

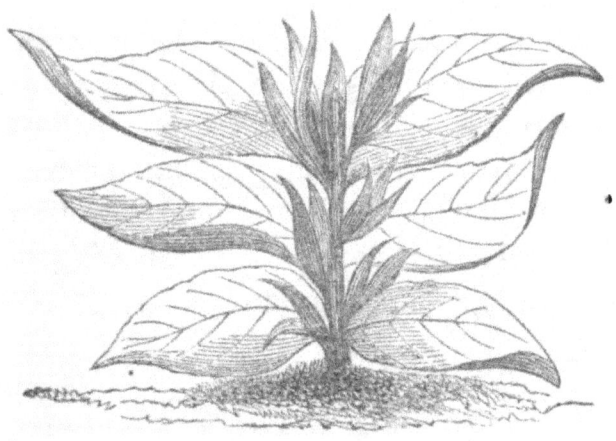

HARVESTING.

Says a Cuban grower, "Tobacco should never be cut before coming to maturity, which is known by the leaves becoming mottled, coarse, and of a thick texture, and gummy to the touch, at which time the end of the leaf, by being doubled, will break short, which it will not do to the same extent when green. It should not be cut in wet weather (nor immediately after a rain, if it can be avoided), when the leaves lose their gummy substance so necessary to be preserved. . . . The grower should be on his guard not to destroy the quality of his tobacco by cutting it too soon.

When the cutting begins, a quantity of forked stakes are set upright with poles thereon to support the tobacco and keep it from the ground. Cut the plants obliquely even with the ground. The person employed should strike the lower end of the stalk of the plant two or three times with the blunt side of his knife, so as to rid it as much as possible from sand and dirt; then tie two stalks together and place them carefully across the poles prepared to receive them. Thus they remain in the sun, or open air, until the leaves are somewhat wilted, so as not to be liable to injury, as when just cut. Then place as many plants on a pole as can be conveniently carried, and remove them to the drying-house, where the tobacco is hung upon the frames prepared to receive it, leaving a small space between the two plants, that air may circulate freely and promote drying. As drying advances, the stalks may be brought nearer together, and thus make room for more." The following cut shows the mode of hanging tobacco on poles.

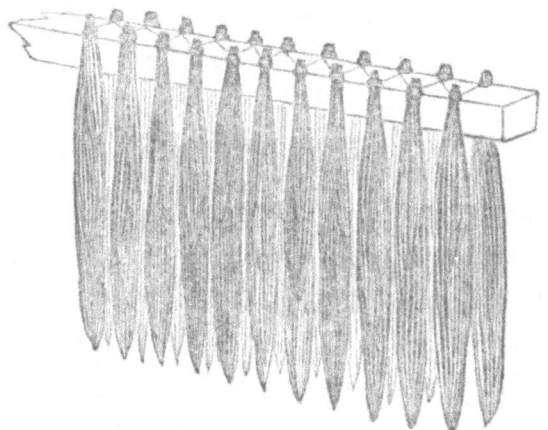

"In drying tobacco," says a Cuban grower, "all damp air should be excluded, as should be drying winds. Drying should be moderately promoted, except in rainy weather, when the sooner the drying is effected the better; for it is a plant easily affected by the changes of the weather after the drying is begun. In damp weather, it is liable to mildew, changing the color of the leaf to a pale yellow, and from this to a brown. When the middle stem is perfectly dry, it can be taken down and the leaves stripped from the stalk and put in bulk to sweat. This is to make tobacco of them; for before this process, when a concentration of its better qualities takes place, the leaves are always liable to be affected by the weather, and cannot well be considered as being anything but common leaves partaking of the nature of tobacco, but not actually tobacco. The leaves are to be stripped from the stalks in damp or cloudy weather, when they are more easily handled and the separation of the different qualities rendered easy. The good leaves are kept by themselves for 'wrappers' or 'caps,' and the defective ones for 'fillings.'"

Paoli Lathrop says, "It will be fit to harvest two or three weeks after topping. Cut it and let it lie on the ground till it is wilted sufficiently to handle without breaking the leaves, avoiding too much exposure to the sun, for sunburning renders the leaf as worthless as if frost-bitten. When removed from the field to the building for curing, it is passed from the load by one man to another, who hangs it by tying the twine round the first plant, and running it over the pole, then, with one turn of the twine, secures every plant till the pole is filled, then fastens the twine. If the pole is twelve feet long, hang from thirteen to fifteen plants on each side and place the poles eighteen inches apart from centre to centre. For the first few days after it is housed, give it plenty of light and air to guard against sweat, which would cause great injury. When all danger from this source is past, keep the building closed, and let the tobacco hang till the stems of the leaves are well cured. It must thus remain until the weather is damp enough to make it soft and pliable. Then it may be cut down by one and passed to another, who packs in a double row, tip to tip, as seen in the following cut:—

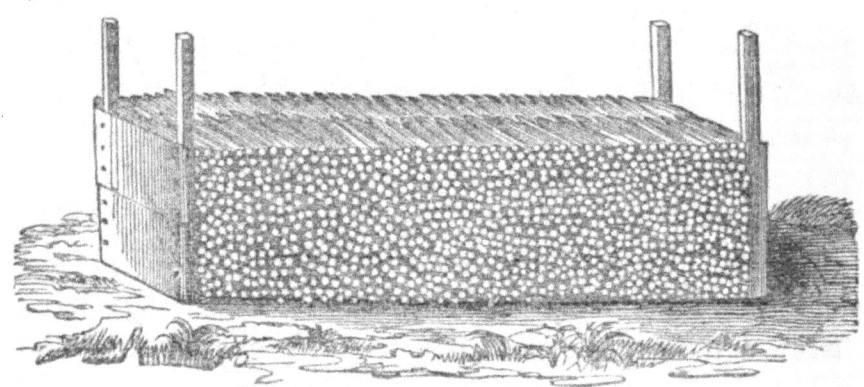

"When thus packed, it should be well covered with straw or cornstalks to prevent drying. Strip it soon after it is taken down, and be careful it does not heat while piled as above. Make two qualities by putting the lower and other poor leaves in hands by themselves." Some make three grades, the best leaves usually being those on the middle of the stalk.

In harvesting, some growers begin when a majority of the plants are ripe, and cut clean as they go; others begin earlier, and cut as it ripens. Both ways have their advantages and their disadvantages. As in this respect, so in others, good growers differ in regard to points of culture and curing, every man whose mind is engaged in his business profiting from his own annual observation and experience.

In stripping, a sufficient number of leaves is tied together to form a "*band*," and the leaves are bent over, forming a head, around which a wrapper is wound and tied. These are laid in piles, the bent ends outward, which, after a few days, will be ready to pack. In Maryland, Virginia, and Kentucky, tobacco is packed in hogsheads, in Massachusetts and Connecticut, in boxes, and thus sent to market.

To recapitulate: The requisites for growing tobacco are, a suitable climate, a good soil, such as will grow from fifty to seventy-five bushels of corn per acre, well cultivated and highly manured, a good plant-bed, thinned and well-weeded, a seasonable transplanting of well-grown plants, clean culture, timely and faithful worming, seasonable topping, prompt suckering, harvesting at the time of full maturity, skill in drying or curing, and in stripping and assorting, so that the leaves shall be free from harmful moisture and in the best possible condition for packing and boxing for market, and sufficient knowledge and forethought to know when, to whom, and for what, to sell the crop.

When all is said that can be gleaned from the best writers and talkers, who publish and state their best observation and richest experience, it is difficult to give a recipe that an unpractised grower can take and follow and be equally succsssful with a skilful and experienced grower of tobacco. The beginner must labor under some disadvantages where every one engaged in cultivating and curing and packing the weed learns something to his advantage every year. Yet it is thought, nevertheless, that the suggestions contained in this MANUAL will serve greatly to aid the inexperienced who desire to engage in growing tobacco, — as many will do when the tobacco crop of an acre of good land sells for nearly six times as much as the corn would bring that might be grown on that same acre, and besides, leave the ground in better condition for any other crop than would the corn.

EXPENSE AND PROFIT OF TOBACCO GROWING.

Elihu Belden, of Whately, Mass., states that he grew on a field of twelve acres, 28,850 lbs. of tobacco, averaging nearly one ton per acre. He stated that he had grown nearly three thousand pounds on an acre. Tobacco has lately sold for 33⅓ cents a pound in Hatfield. At that rate, the tobacco crop of an acre, (three thousand pounds), would sell for $1000; of an acre, (two thousand pounds), it would sell for $666,66; of an acre, (fifteen hundred pounds), it would bring $500. Mr. Belden stated that the cost per acre in growing the twelve acres, producing as aforesaid, including the interest on land at $100 per acre, manure at $1.50 per load, guano at three cents a pound, superphosphate at two and a half cents a pound, and labor, was $109.50. The cost of growing an acre in Onondaga County, N. Y., five years ago, was set down as follows: Expenses,—plants, $2.50; manure ten cords, $20; fitting ground and working, $4.50; transplanting, $5.00; cultivating and first hoeing, $2.00; cultivating and second hoeing, $1.50; topping and worming, $1.00; suckering three times, $6; harvesting and hanging, $6.00; stripping one ton, $10.00; five packing boxes, $5.00; labor of packing, $1.50; twine for hanging, $1.00; aggregate, $66. Returns: two thousand pounds of tobacco at 13 cents a pound, $270; deduct for shrinkage, transportation, and commission, $52; which, taken from the aggregate, leaves $218; from which take the expense of growing, leaves a net profit of $150 for one acre. In Connecticut, 2500 pounds per acre have been grown on fields of many acres in extent, the grower realizing $400 per acre for his crop. Such results are due to superior soil and culture.

It is a question with some whether high culture is more profitable than the "four-shift system" of Maryland and

Virginia, with clover as its means of fertilization. With the tobacco growers, in the valley of the Connecticut this is no question. Guano and artificial fertilizers are now used South as here. O. W. Bryan of Maryland in his prize essay on tobacco growing, recommends "Peruvian Guano, hog-manure, well-rotted oak ashes, well-rotted stable-manure, with plaster, — guano at the rate of one thousand pounds per acre." W. W. Bowie, of Maryland, in his prize essay, advocated a liberal top-dressing every ten days of a compost of "unleached ashes, pulverized sulphur, plaster, and salt, — would apply guano on light soil, but not on rich land." Substitute for these commercial manures, or fertilizers, Bradley's TOBACCO FERTILIZER. In Maryland, Kentucky, and Ohio, crops are as to quality of soil; in the latter two States, as reported, the crop has averaged from one thousand pounds to two thousand pounds per acre.

A writer stigmatizes tobacco-growing as the bane of Virginia agriculture. It is the mode of cultivation, rather than the kind of crop grown, that has impoverished the soil of the Old Dominion. The same deterioration of soil would have been true of any other crop as persistently grown, with shallow culture and no manure, as of tobacco. The tobacco farms in the Connecticut Valley have constantly improved under the culture of the weed. Said one farmer, "When I began to grow tobacco, I could keep but about twenty head of cattle; now I keep nearly one hundred." Another said, "Where I cut less than a ton of hay per acre, before I began to grow tobacco, I now cut over three tons of good hay per acre."

TOBACCO-BARNS, OR DRYING-HOUSES.

Tobacco-barns, drying-houses, or sheds, constitute a very essential and important part of the outfit of tobacco-growers. It is deemed important, therefore, to devote some space to the consideration of the subject. The first cut below is of a barn one-half open on a part of the side and the entire end, in order to show the tobacco as hung up to cure. The further portion of the side view, represents every third board hung on strap-hinges.

The second cut represents a barn open at the bottom, two feet from the sill all round, and open at the top, also, as represented in the cut below, so as to favor ventilation, which, it is maintained by some, is quite sufficient for curing the weed after it is hung on poles. Others doubt this, though it is adopted and approved by some successful planters who have tried it. A barn thirty-six feet by twenty-four, with twelve-foot posts, will be sufficient for an acre of tobacco, say some growers.

21

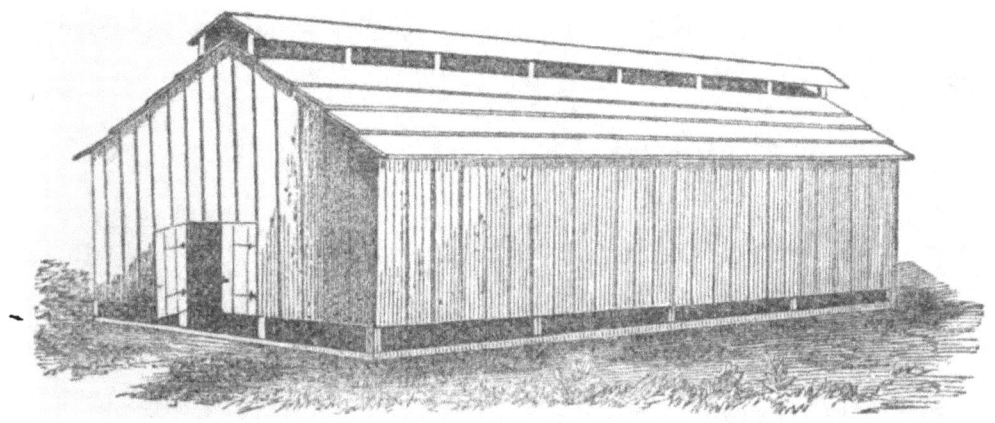

The cut below was taken from a drawing made by L. B. Field, of Hatfield, a builder of tobacco-barns in that and neighboring towns. The first cut below shows the framework of a barn sixty feet by thirty-nine feet.

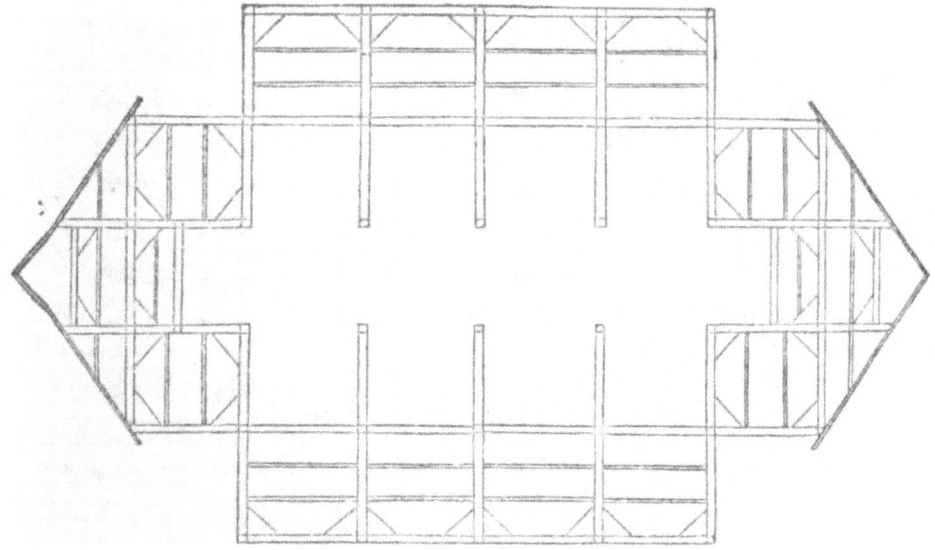

BILL OF TIMBER AND LUMBER.

The following bill of timber and boards for a barn like the above, thirty-nine by sixty feet, was furnished by Mr. L. B. Field, giving length and dimensions of the timber for the frame, with the kind of boards, and the quantity of the same, and shingles, for enclosing such a barn.

```
 8 sills 15 feet long, 7 in. by 7
10 do.  13¼ do.      7  do. 7
 5 beams 39½ do.     7  do. 7
20 posts 14  do.     7  do. 7
 8 plates 15 do.     7  do. 7
 8 P. pl't 15 do.    7  do. 7
10 P. posts 8¼ do.   7  do. 7
10 girts 13  do.     6  do. 6
32  do.  15  do.     6  do. 6
16  do.  15  do.     4  do. 4
12  do.  13  do.     4  do. 4
 4  do.   8  do.     4  do. 4
80 braces 4¾ do.     3  do. 4
 6  do.   4  do.     3  do. 4
48 rafters 24½ do.   3  do. 4  - - - -  8281
484 poles 13 do.     2  do. 6  - - - -  6292
   Boards 14½ do. - - - - - - - - -    2900 ft.
     do.  14 do. for gable ends, - - - - 560 "
   Roof-boards, - - - - - - - - -      2430 "
   Lining-boards, 4 inches in width, - - 1500 "
   Finishing boards, - - - - - - - -    250 "
   Shingle, if laid 5½ in. to weather, 21,000
                                      ─────────
                                       22,213 "
```

The second cut represents the above frame as raised and enclosed. The builder, Mr. Field, instead of hanging every third board on hinges, hangs them thus: three boards in a section of fifteen feet,—that is to say, one board is hung so as to open in every five feet of the enclosing, with an opening in each roof, near the ridge-pole, or where it would be, if there were one, to promote ventilation, as in the barn above, represented as open only at the bottom and the top. In constructing the opening in each roof, care should be taken to prevent the water from driving in, when it rains, as this may easily be done.

A MODEL TOBACCO-BARN.

The tobacco should not be hung too near the ground in the barn or drying-house. If the building is on wet or moist, tenacious ground, it should be thoroughly underdrained, as this will favor the drying of the weed, and also prevent the liability to mildew in wet weather. Always be careful to hang the plants so that there shall be free circulation of air between the poles upon which the tobacco is suspended to dry. Every one must be his own judge, as the distance apart in hanging to dry must depend upon the size of the plants to a certain extent.

CONSUMPTION AND EXPORTATION OF TOBACCO.

THE number of persons employed in growing, curing, and manufacturing tobacco in the United States is not given in the census report. In the city of Hamburg, Germany, the manufacturing of tobacco gives employment to more than 10,000 persons, who turn out 150,000,000 cigars a year, valued at $2,000,000. From Havana and Manilla, Hamburg imports 18,000,000 cigars a year, making an aggregate, including its own production, of 168,000,000 cigars, 153,000,000 of which are exported, leaving 15,000,000 for home consumption; allowing 40,000 cigars a day to an adult male population of 45,000. In England, with a population of 21,000,000, in 1821, the consumption of tobacco was 15,598,152 lbs., an average of 12 ounces per head for the entire population; in 1831, with a population of 24,410,439, the consumption reached 19,533,841 lbs., or 13 ounces per head; in 1841, population 27,019,672, consumption, 22,309,360 lbs., or 13½ ounces per head; and in 1851, population 27,452,692, the consumption was 28,062,841 lbs., or 17 ounces of tobacco per head, showing a steady increase. In France the consumption of tobacco is 18½ ounces per head, nearly half of which is snuffed; in Denmark, in 1848, it was 70 ounces per head; and in Belgium, it averages about 73½ ounces per head. A popular writer sets down the consumption of tobacco by the whole human family, annually, at 2,000,000 tons, or 4,480,000,000 lbs., or 70 ounces per head; and he adds that "the annual tobacco crop of the world weighs as much as the wheat consumed by 10,000,000 of Englishmen, with a money value equal to all the wheat consumed in Great Britain."

Tobacco is called an exhausting crop. So is any crop when annually exported and nothing returned to replace the elements of soil thus carried off. The exportation of breadstuffs of all kinds, cotton, tobacco, etc., must impoverish the soil of the country where grown, and enrich that where they are received. Farmers thus make a slight profit, the middle men a large gain; leaving the prospect for future generations anything but encouraging. It is bad economy to exchange the produce of the soil of one nation for the manufactured articles of another, unless an amount of fertilizers equal to what is exported be imported. It is better for the farmer, and, therefore, for the country, to give him the opportunity of feeding the English operatives at home than to furnish half enough to feed them in Manchester. In view of this doctrine, no farmer can fail to see that he impoverishes his farm by the annual exportation of his tobacco crop. Hence, he cannot fail to see that if he can import and use to advantage a commercial fertilizer, he should do it. It is claimed — and this claim is confirmed by ample testimony, as may be seen in perusing the numerous testimonials contained in this Manual — that Bradley's Patent Tobacco Fertilizer is just what the tobacco-grower needs to replace, in a measure, the mineral elements removed by the crop.

ANALYSIS OF TOBACCO SOILS AND TOBACCO.

AT the request of the Commissioner of Patents, Dr. Charles T. Jackson, of Boston, Chemist and Geologist, obtained specimens of surface and sub-soils from tobacco-growing districts of Maryland, and the Connecticut River Valley, in Massachusetts, a few years since, and submitted them to a thorough analysis, in order to ascertain the proportions of the elements in the soils appropriated by the plants grown thereon as shown by their analysis. The samples of soils were obtained by driving tinned iron tubes, 20 inches in length, 2 inches in diameter, and 20 inches long, into the ground, so as to cut out a sample of the soil to the depth of 20 inches. The fields from which the samples were obtained in Maryland are in Prince George's County, and had been cultivated many years with red clover, ploughed under, alternated with crops of tobacco and grain. The Massachusetts samples were obtained from Hatfield and Whately, about 30 feet above the level of the Connecticut River, from fields not subject to being overflowed, and that have been manured from the barn-yards. The Maryland soil contained a larger proportion of magnesia than the Connecticut River alluvion; hence the larger proportion of magnesia in the Maryland tobacco, the lime salts appearing to take the place of the magnesian in the Massachusetts tobacco.

By the analysis of the ash of plants, knowledge is obtained in regard to the mineral matters which plants appropriate from the soil. It was found that 1000 grains of the Hatfield soil yielded to a solution of carbonate of ammonia 0.39 grains of the phosphates; the same quantity of Whately soil, 0.587 grains, and the Maryland soil, 0.85 grains. Carbonate of ammonia furnishes nitrogen and other requisite saline matters. The quality of tobacco is largely the result of the kind of soil in which it is grown. The green plant contains 88 per cent. of water, charged with saline matter, so that when 100 lbs. of tobacco are dried, they weigh but 12 lbs., the salts of the 88 lbs. of juice being concentrated in the 12 lbs. of dried leaves.

The analyses of the ash of tobacco plants from Maryland and Massachusetts were made to determine what mineral elements are taken from the soils by the crops, so that the growers may learn what fertilizer they shall use in order to replace them. It is admitted that the tobacco plant is a voracious feeder upon soil-elements essential to its growth.

It appears that animal manures, furnishing nitrogen, cause the plant to grow rank, while the flavor is materially impaired, and a much larger proportion of nitre is introduced into the plants, so that those grown on richly-manured and old soils will burn with a decrepitation, or crackling noise, like salt when heated, or saltpetre paper. This is more manifest in the leaf-stems than other portions of the plant. Of the Hatfield tobacco 100 lbs. dried, were reduced to 11 lbs. It is remarked that the best Cuba tobacco does not deflagrate or burn with decrepitation, while samples from the Connecticut River and from Manilla are remarkable for vivid decrepitation like saltpetre paper, owing to the presence of nitre. The nitrates contained in tobacco affect its combustion. The stalks of the plants are richest in the alkalies, potash, and soda, while the leaves contain a proportion of lime. Magnesia and lime replace each other in a curious manner, which is owing to the nature of the soil.

From what is learned by analyses, potash, soda, lime magnesia, and phosphoric acid seem to be the essential elements of fertility in tobacco soils; for these are the substances which are most largely removed by the crop. Silica, oxyde of iron, and sulphuric acid are rarely wanting, and chlorine is in most localities abundant in the condition of sea-salt and muriate of lime. Organic manures seem to be unfit for this crop; at least those of a nitrogenous, or ammoniacal character, tend to produce too much saltpetre, causing the plant to grow rank and coarse.

Dr. Jackson also adds that animal manures and such as give a considerable quantity of soluble organic matter which is absorbed by the plants, tend to produce too rank a growth so that the tobacco has not so fine a flavor as when it is less liberally supplied with soluble organic compounds.

See table of analyses on the following page:—

ANALYSES OF TOBACCO PLANTS FROM MASSACHUSETTS AND MARYLAND.

BY DR. CHARLES T. JACKSON.

Names of Ingredients.	Locality, Massachusetts — Hatfield, Connecticut River, farm of W. H. Dickinson. Sample from the best soil. Per cent. of leaf.	Locality, Massachusetts — Town of Hatfield, Connecticut River, farm of W. H. Dickinson. Sample from best soil. Per cent. of stalk.	Locality, Massachusetts — Town of Whately, Connecticut River, farm of J. Allis. Tolerably good soil. Per cent. of leaf.	Locality, Maryland — Prince George's county. Richest soil. Per cent. of leaf.	Locality, Maryland — Prince George's county. Richest soil. Per cent. of stalk.	Locality, Maryland — Prince George's county. Much worn soil. Per cent. of leaf.	Locality, Maryland — Prince George's county. Much worn soil. Per cent. of stalk.
Silica and silicious dust	9.60	0.40	29.40	8.60	2.40	21.20	3.20
Phosphoric acid	7.60	12.52	9.05	8.50	12.52	7.15	10.28
Lime	25.75	11.84	28.99	22.66	11.48	25.85	23.88
Potash	20.40	40.12	15.20	17.60	40.12	20.32	27.48
Soda	6.03	9.20	2.52	1.40	9.20	4.36	7.28
Magnesia	1.60	0.80	0.60	8.00	0.80	2.00	0.40
Peroxide of iron and manganese	1.20	2.00	1.60	2.80	1.40	1.20	1.30
Chlorine	1.68	2.96	0.72	3.76	2.96	0.92	3.12
Sulphuric acid	2.75	2.04	2.72	8.00	2.04	1.52	4.48
Carbonic acid	21.20	16.00	9.20	18.40	16.00	14.80	18.00
Loss	2.19	2.12	0.00	0.28	1.08	0.68	0.68
	100.00	100.00	100.00	100.00	100.00	100.00	100.00
	Per cent. of ash. 18.82	Per cent. of ash. 16.73	Per cent. of ash. 20.3	Per cent. of ash. 14.08	Per cent. of ash. 9.3	Per cent. of ash. 14.76	Per cent. of ash. 8.72

TESTIMONIALS.

WEST SPRINGFIELD, MASS., Dec. 22d, 1863.

To WM. L. BRADLEY. — Dear Sir: We used your "Patent Tobacco Fertilizer" on our tobacco crop the past season with most excellent success; it gave the plants an immediate start and vigorous growth to the end, and thus insured a splendid crop.

        Yours truly,   RICHARD COOLEY,
                                WALTER COOLEY.

HATFIELD, MASS., Dec. 23d, 1863.

To WM. L. BRADLEY. — Dear Sir: I procured a few bags of your "Tobacco Fertilizer" the past season and used it on a part of my tobacco crop, in order to determine for myself its value as a fertilizer for growing tobacco. I can therefore most cordially recommend its use to all tobacco growers. It did remarkably well for me in every trial. I have never used anything equal to your "Patent Tobacco Fertilizer," for giving plants an early and rapid growth.   Truly yours,

                JONATHAN S. GRAVES.

HATFIELD, Dec. 22d, 1863.

To WM. L. BRADLEY. — Dear Sir: Having had of late but little confidence in patent manures, yet I was persuaded, when setting my tobacco, to purchase some of your "Patent Tobacco Fertilizer," in order to make one more trial, and I must say, that I liked the article much; for it starts the young plants quick, gives them a dark color and rapid growth; make it as good as it was last season, and it is *the* thing needed for tobacco.

    Truly yours, EDWIN BRAINARD.

HATFIELD, MASS., Dec. 24th, 1863.

To WM. L. BRADLEY. — Dear Sir: We used your "Patent Tobacco Fertilizer" on a portion of our tobacco fields with most excellent results. We cheerfully recommend it as a valuable fertilizer. We saw its effects on many of our neighbors' tobacco fields, and in all cases it did well. We would advise the use of a liberal handful in every hill.

    Very truly yours, J. F. & G. C. FITCH.

SHELBURNE FALLS, Nov. 23d, 1863.

To WM. L. BRADLEY. — Dear Sir: For the benefit of farmers, I desire to state, that having used Bradley's "Tobacco Fertilizer" the past season, so well pleased was I that I shall use it in the future. I applied it to little less than an acre of exhausted meadow, — having cut thereon in the last week of June, not exceeding half a ton of hay per acre. I ploughed the same the first week in July, topdressed with about 20 loads of compost manure, harrowed thoroughly, and set it to tobacco on the 12th of July, with a handful of the "Patent Tobacco Fertilizer" in the hill. The result was about a ton of fine tobacco, exeeeding fully one-half my expectation, which was, no doubt, the effect of the fertilizer.

    Yours, SAMUEL D. BARDWELL.

WEST SPRINGFIELD, MASS., Jan. 4, 1864.

To WM. L. BRADLEY. — Dear Sir: The parties using your "Patent Tobacco Fertilizer" on my grounds the past season give it their unqualified approbation.

    Respectfully Yours, JUSTIN ELY.

WALLINGFORD, Ct., Dec. 26th, 1863.

To WM. L. BRADLEY.—Dear Sir: I used your "Patent Tobacco Fertilizer" last season on my tobacco fields, and pronounce it a tip-top article, as good a fertilizer as can be, and shall use it freely another season.

Yours truly, GOULD N. ANDREWS.

NORTHAMPTON, Dec. 3d, 1863.

To WM. L. BRADLEY.—Dear Sir: We send you Mr. Thayer's statement about the "Patent Tobacco Fertilizer:" it is voluntary, and accords with the universal testimony of all to whom we sold it so far as we have heard.

STOCKWELL & SPAULDING.

NORTHAMPTON, MASS., Dec. 3d, 1863.

To WM. L. BRADLEY.—Dear Sir: I tried your "Patent Tobacco Fertilizer" on my tobacco. I staked out the rows where it was applied so as not to be mistaken, and I do not hesitate to say, it was the best fertilizer I used last season. I also regard it as the cheapest fertilizer in use to bring the crop forward early.

Truly yours, JUSTIN THAYER.

HATFIELD, MASS., Dec. 24, 1863.

To WM. L. BRADLEY.—Dear Sir: Your agent called on me last Spring to sell me some of your "Patent Tobacco Fertilizer." I told him I had but little faith in it, having used different brands of fertilizers for several years, and found them deceptive. He insisted that I should try one bag. I said I would not buy one pound, but if he desired, he could leave some and I would experiment with it if he wished. He did so, and now for the result. I tried it on tobacco by the side of Peruvian Guano, ploughed under my manure, and put the fertilizer in the hill. I tried it on three rows, put the fertilizer on the outside one, not giving it an equal chance with the guano, but it started first and kept ahead through the season. It was plain to be seen that it was heaviest when harvested. Hence I conclude that Bradley's "Patent Tobacco Fertilizer" is a most valuable and genuine article. I shall use it next season with confidence.

   Truly yours,   ELISHA MARSH.

WALLINGFORD, Ct., Dec. 23, 1863.

To WM. L. BRADLEY.—Dear Sir: I raised my first tobacco crop last season, and hoping to make my first attempt a success, I prepared a composition which I thought to be exceedingly rich, made of my finest and best barnyard manure and other valuable substances common on every farmer's premises. I first manured my field thoroughly and applied my compost in the hill. For experiment I obtained a small quantity of your "Patent Tobacco Fertilizer" and applied it on a few rows in the hill, where no compost was used. The result was, that when the tobacco set upon your Fertilizer was knee-high I could not see at forty rods distance that there was any tobacco on the remainder of the field, all of which had received a liberal application of my compost, that I thought could not be beaten. The cut worm worked badly except where your Fertilizer was used; and in fact I was compelled to set over so many times that I became nearly discouraged. I think no plant set on your "Tobacco Fertilizer" was attacked by the cut worm. The tobacco grown on your "Fertilizer" ripened two weeks earlier than any other part of the field, which establishes in my mind its superiority as a fertilizer for the tobacco crop. Respectfully Yours,

          WM. W. IVES.

WHATELY, Dec. 23, 1863.

To WM. L. BRADLEY. — Dear Sir: Experiments with your "Patent Tobacco Fertilizer" the past season convinced us that it is truly a valuable article for the cultivators of tobacco. We used it on various soils and in every trial it gave remarkable results. It started the plants earlier than anything we used, and made a vigorous growth to maturity. We ploughed an exceedingly poor knoll (the poorest piece of ground, we thought, on our farm) late in the season, and set it to tobacco, using no manure but your "Patent Tobacco Fertilizer," and we never had tobacco grow more rapidly than this, although we did not put a hoe into it at all and got a splendid crop. We shall use a liberal quantity of your "Tobacco Fertilizer" on all our crop next season.

Yours truly, J. B. MORTON & SON.

WALLINGFORD, CT., Dec. 29th, 1863.

To WM. L. BRADLEY. — I tried your "Patent Tobacco Fertilizer" last season with other fertilizers, and found nothing superior to it; and being well pleased with the result, I do not hesitate to recommend it as a valuable fertilizer. Truly yours, WM. FRANCIS.

WEST SPRINGFIELD, MASS., Dec. 30, 1863.

To WM. L. BRADLEY. — Dear Sir: I purchased a small quantity of your "Patent Tobacco Fertilizer" last season and gave it a thorough trial; and I am so well pleased with it that I shall use no special fertilizer hereafter on my tobacco crop but your "Patent Tobacco Fertilizer." It gave the largest yield of anything I applied, though used on the poorest part of my field. The plants set upon your fertilizer grew more rapidly, and gave larger and broader leaves. It is most valuable for plants set late, as it makes them grow as by magic, and continues to act to the time of ripening.

Very truly yours, GEO. N. HALL.

HATFIELD, Dec. 29th, 1863.

To WM. L. BRADLEY. — I purchased one bag of your "Patent Tobacco Fertilizer" last spring, applied it to part of my tobacco field, at the rate of about two table-spoonfuls in a hill. It made a decided increase in the quantity of tobacco. My field was all well manured. The first month of the growth the tobacco set upon your fertilizer was as large again as that without it, — an important result, as thereby the crop is brought to maturity much earlier. Truly yours, J. E. WIGHT.

HATFIELD, MASS., Dec. 3d, 1863.

To WM. L. BRADLEY. — Dear Sir: I bought of your agent here, Mr. J. G. Dickinson, a few bags of your "Patent Tobacco Fertilizer" and used it on some of my tobacco with great satisfaction. It started the tobacco quickly. I think it a valuable fertilizer, and shall use a much larger quantity next season, and trust it will be supplied in sufficient quantities to accommodate all customers.

Respectfully Yours, JAS. W. WARNER.

NORTHAMPTON, MASS., Dec. 3d, 1863.

To WM. L. BRADLEY. — Dear Sir: I can recommend your "Patent Tobacco Fertilizer" as a most valuable manure for growing tobacco, as it gives an immediate and rapid growth. I shall use a large quantity for my tobacco another season.

Very truly, &c., JOHN B. GRAVES.

HARTFORD, CT., Jan. 9th, 1864.

To WM. L. BRADLEY. — Dear Sir: I used your "Fertilizer" last season on a part of my tobacco field with splendid success. About one-half of my ground was highly manured with the best stable and hog-pen manures and was thoroughly cultivated, it being an old garden spot. This I thought rich enough without special fertilizers.

On the other part of my ground I cut about the last of June a fair crop of grass (it had not been ploughed for several years), turned it over flat, scattered on broadcast at the rate of six or seven hundred pounds of your "Fertilizer" per acre and harrowed thoroughly. On the first of July I set my plants in the usual manner (the other ground had been set two weeks), with a small quantity of your "Fertilizer" in the hill. The plants grew astonishingly fast from the setting.

I applied the fertilizer around and near the plants at the first and second hoeings. From careful observation I am decidedly of the opinion that the fertilizer should be applied at least twice during the early growth of the tobacco and there is no more favorable time than at the two hoeings, as it is desirable that it should be mixed with the soil near the roots of the plants. This was the manner of my treatment, and I am confident that the last applications were quite as beneficial as those previous to the setting.

Many of my friends called to look at my experiment with your "Fertilizer" in comparison with the manure, and they all said that the tobacco grown on your "Fertilizer" was the best yield. Its color was much the darkest, the growth the most rapid, and it ripened about the same time with that planted two weeks earlier. The yield was estimated at 2000 lbs. per acre; but as I have not stripped it, I am unable to give the result in figures. I think from my experience that quite as large a crop, and a decidedly finer quality, of tobacco can be grown by the proper use of your "Fertilizer" as upon stable manure.

Very truly yours, W. R. LOOMIS.

WEST MERIDEN, CT., Jan. 6th, 1864.

TO WM. L. BRADLEY.—My Dear Sir: I applied the "Patent Tobacco Fertilizer" I had of you the past season on a portion of my tobacco in the hill, and carefully noted its effects. I have no figures by which to give exact results. But it fully met my expectations upon the newly set plants, and I do not hesitate to recommend it as a valuable fertilizer for tobacco.

Truly yours, OLIVER RICE.

DIRECTIONS FOR USING BRADLEY'S PATENT TOBACCO FERTILIZER.

When the land has received a liberal quantity of barn-yard or hog manure, use a handful of the Fertilizer to each plant at the time of setting, mixing it well with the soil.

When the land has received but little manure, use a handful of the Fertilizer to each plant when setting, mixing it well with the soil; and a handful to two or three hills, scattered round the plants at the second hoeing, covering it slightly with earth.

When the grower would depend *entirely* upon the Fertilizer for raising a crop of tobacco, use from 600 to 800 lbs. of the Fertilizer per acre, broad-cast, harrow it in *thoroughly*, and use a handful of the Fertilizer to each plant when setting, mixing it well with the soil.

NOTE.—It is recommended by some to raise the earth a little above the level of the ground, where the plants are set.